Nanami

Daiki

Yui

Hiroto

Let's learn mathematics together!

Grade 1 Vol.2

5

1 Numbers up to 10

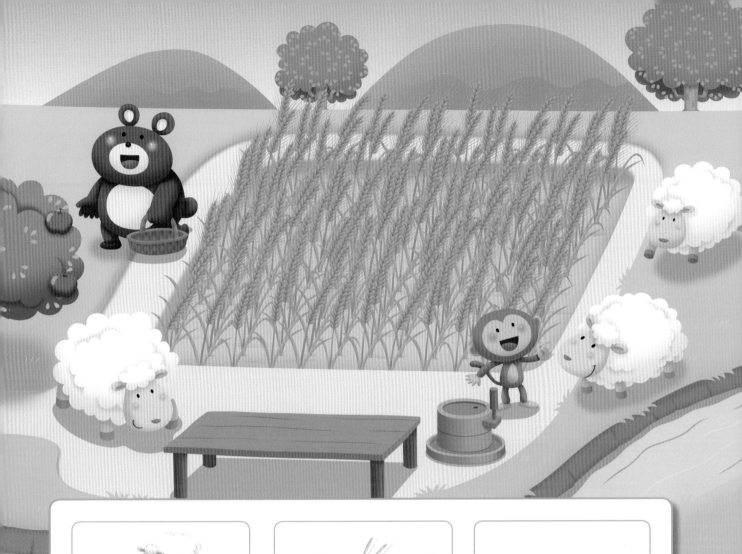

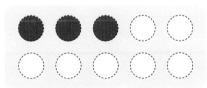

3

three

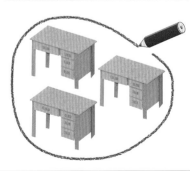 

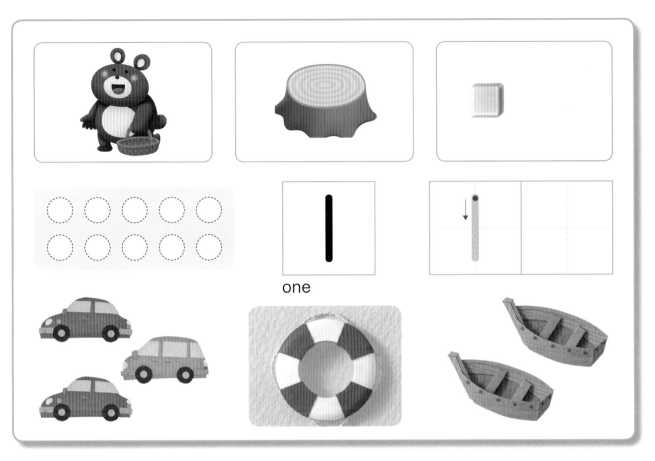

one

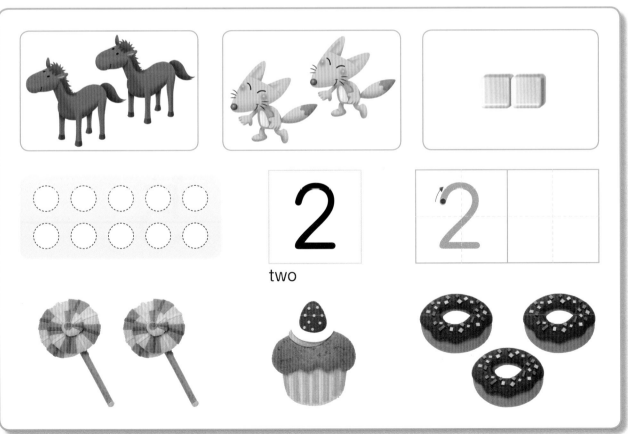

two

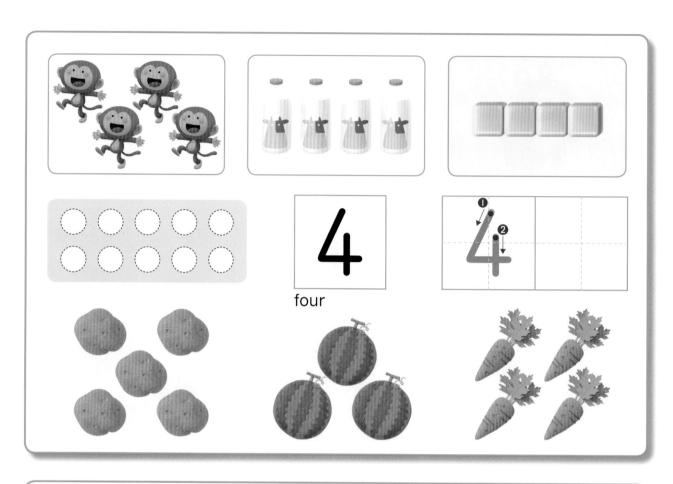

four

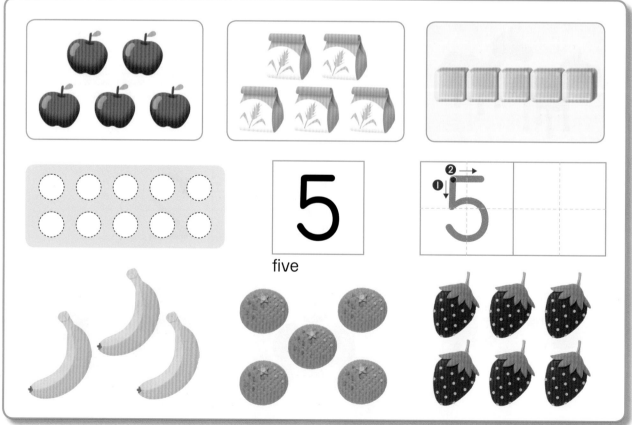

five

  two

three

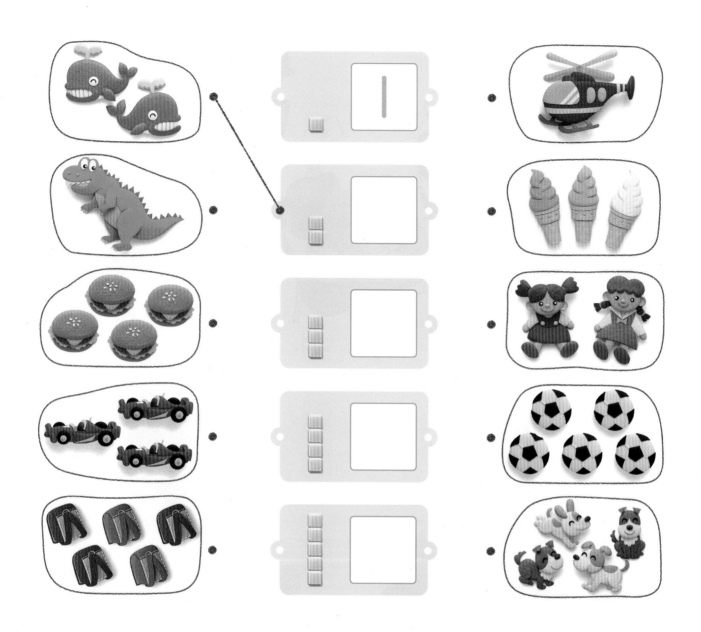

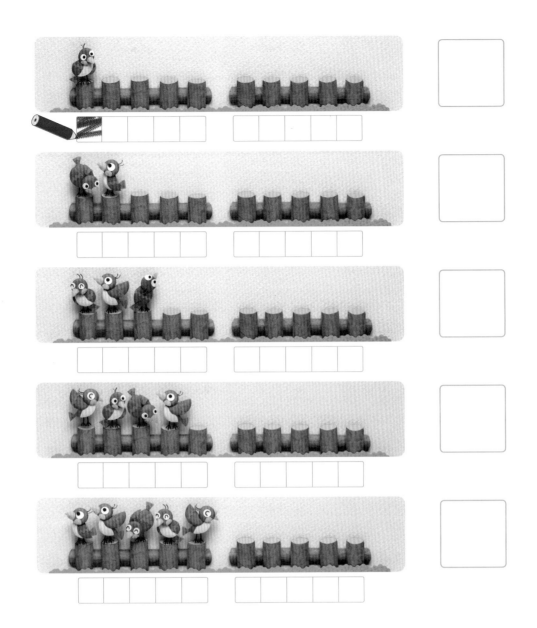

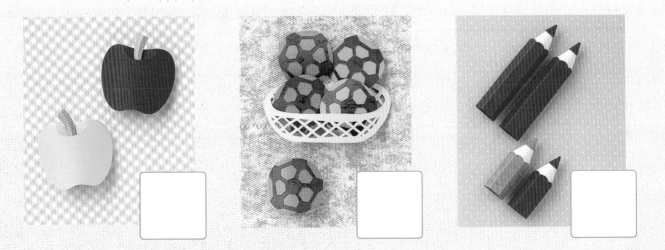

six

seven

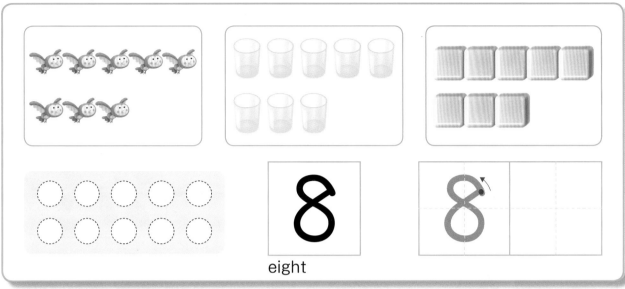

eight

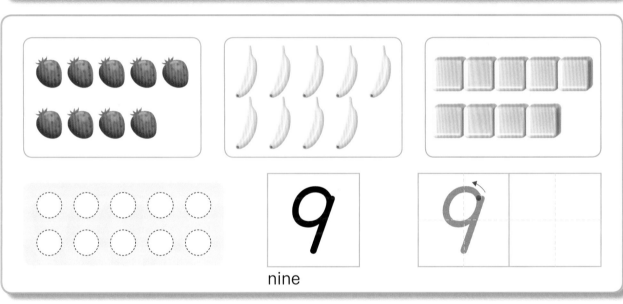

nine

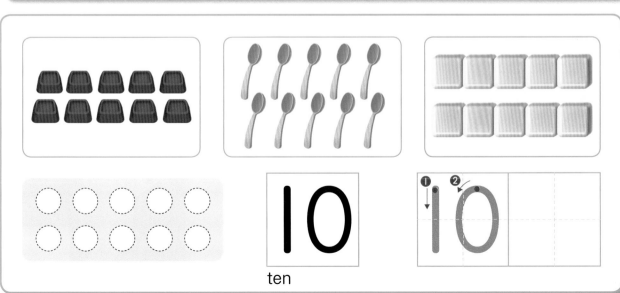

ten

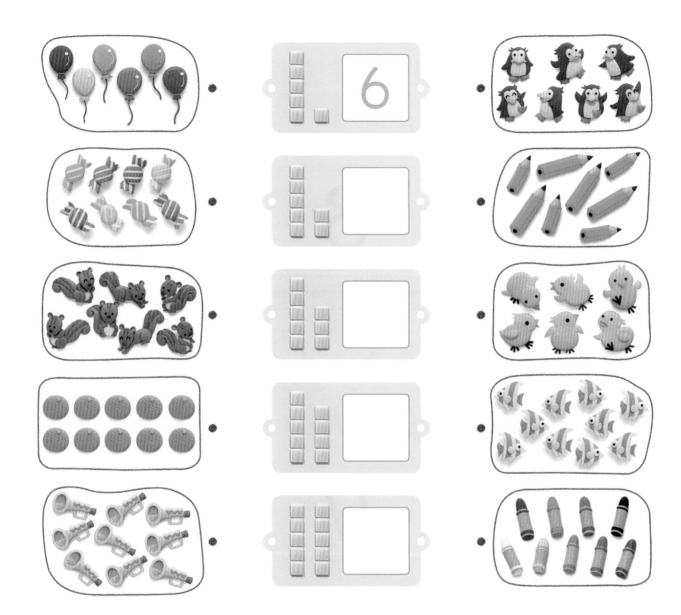

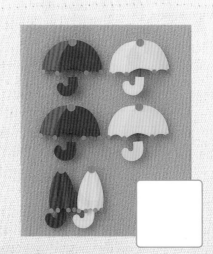

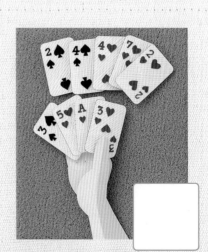

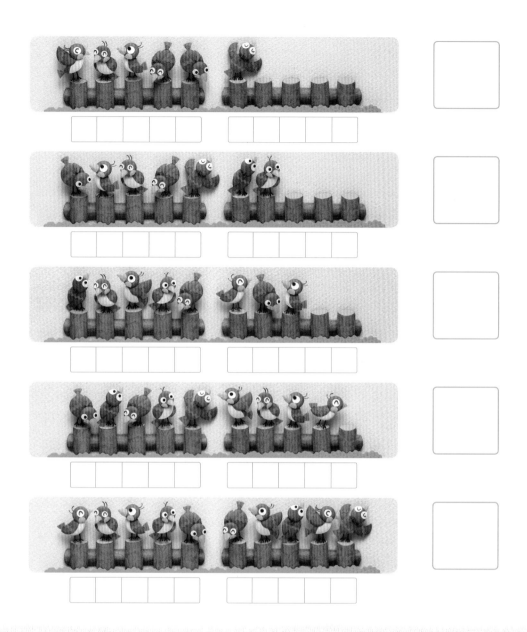

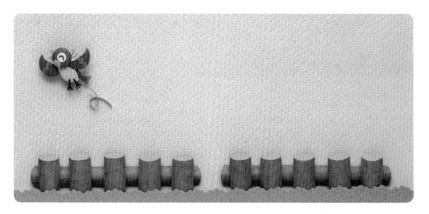

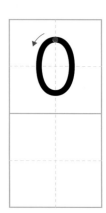

zero

How many rings will enter?

Which has more?

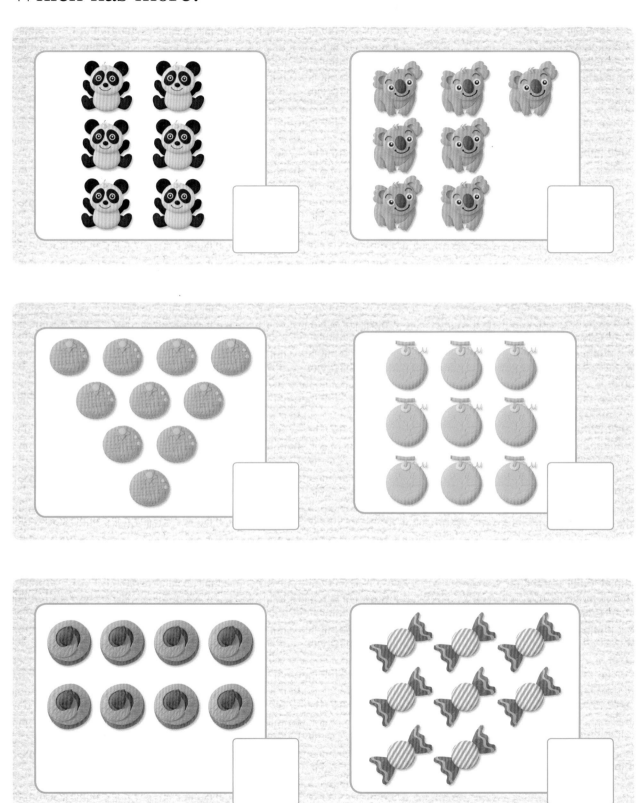

Which is larger?

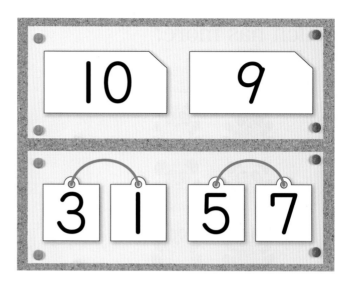

Ordering the cards

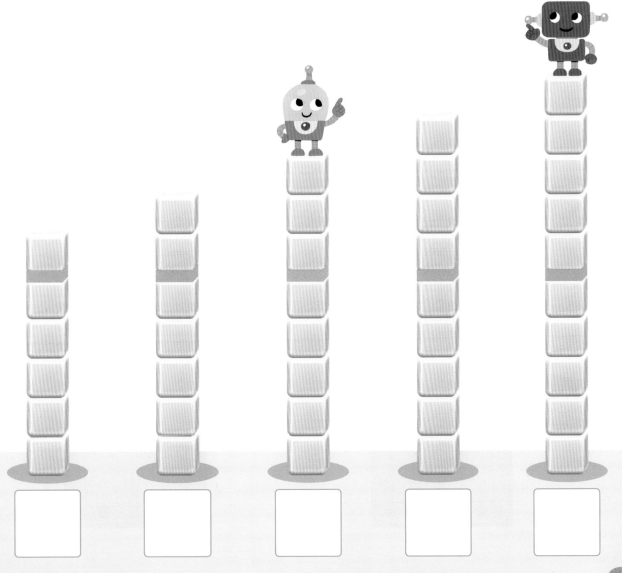

2 Decomposing and Composing Numbers

Let's try to put
5 balls into the box.

Daiki

Yui

Hiroto
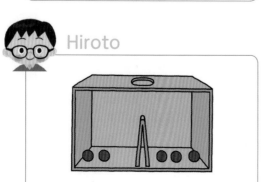

Nanami

5

5 balls are divided into two sides. How many balls are there in each side?

Let's write the number in the ▢ .

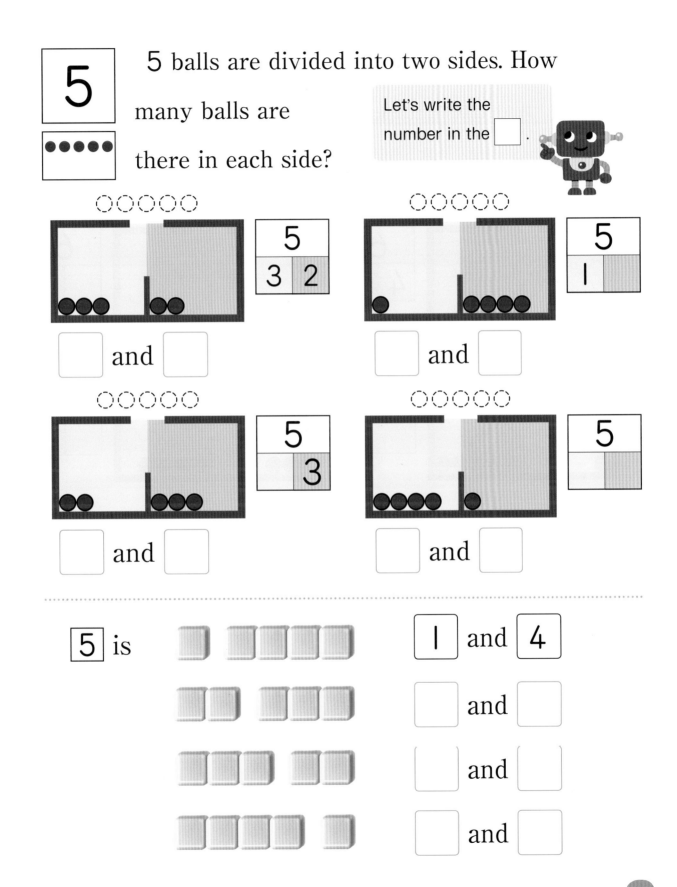

5
3 2

▢ and ▢

5
1

▢ and ▢

5
  3

▢ and ▢

5

▢ and ▢

5 is

1 and 4

▢ and ▢

▢ and ▢

▢ and ▢

6 balls are divided. How many balls are there in each side?

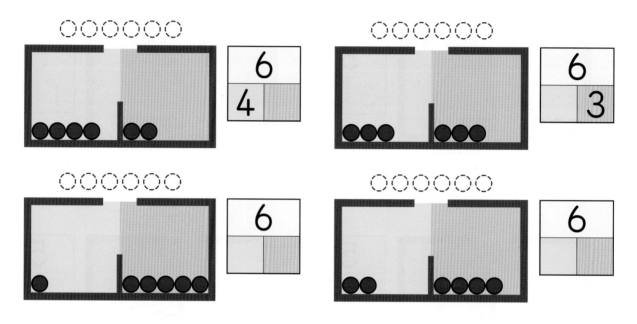

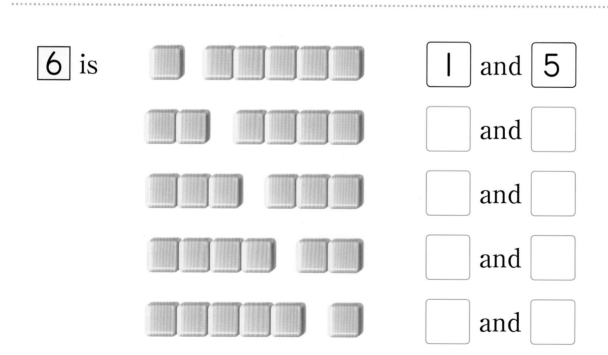

6 is

1 and 5

☐ and ☐

☐ and ☐

☐ and ☐

☐ and ☐

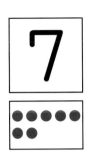

I got **7** counters.

7	
	5

7	

7	

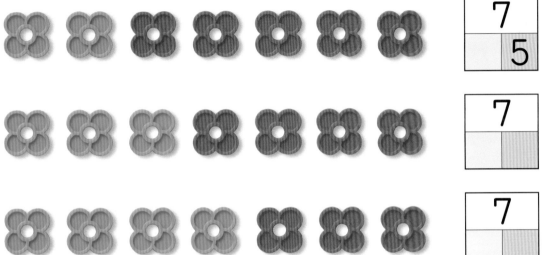

7 is

| 1 and 6 |
| □ and □ |
| □ and □ |
| □ and □ |
| □ and □ |
| □ and □ |

I have 8 counters.

How many counters are hidden?

8	
3	5

8	
2	

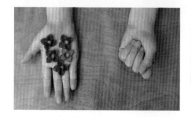

8	
5	

8	
	4

8 is 1 and 7

and

and

and

and

and

and

9 Let's make 9 by using two cards.

5   1   3   6   7

6   4   8   2   3

9 is ... 1 and 8

... and

... and

... and

... and

... and

... and

... and

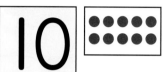

Which pairs are
similar to each other?

1	and	9	10

	and		10

	and		10

	and		10

	and		10

	and		10

	and		10

	and		10

	and		10

Let's make 10

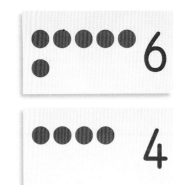

1 If you make 10, show your cards.

Will these make 10?

5 and 5 make 10.

Let's circle two numbers which make 10.

5	8	2	6
5	1	3	4
8	9	7	5
4	6	5	7

2	6	4	3
1	8	7	8
9	6	5	2
4	7	3	5

Let's find pairs which are arranged vertically, horizontally and diagonally.

3 How Many : Altogether and Increase

Let's look at the pictures and tell a story.

①

I have ☐ balls.

I have ☐ balls.

→ We have ☐ balls altogether.

②

I have ☐ blocks.

I have ☐ block.

→ I have ☐ blocks altogether.

1 How many goldfish are there altogether?

Let's think about the problem by using blocks.

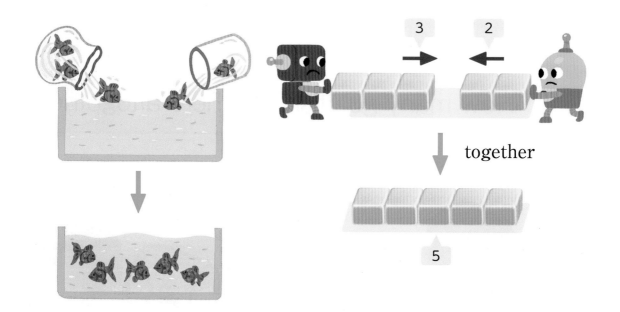

Putting 3 and 2 together makes 5.

Math Sentence : 3 + 2 = 5   Answer : 5 goldfish

3 plus 2 equals 5

 3 + 2 = 5 is a math sentence,
3 + 2 is a math expression.

2 How many frogs are there altogether?

Math Sentence :

2 + 1 =

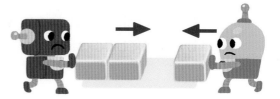

Answer : ☐ frogs

1 ▶ How many children are there altogether?

Math Sentence :

☐ + ☐ = ☐   Answer : ☐ children

2 ▶ Let's do **addition**.

2 + 2   1 + 4   3 + 1   1 + 2

2 + 1   2 + 3   1 + 1   1 + 3

3 There are 5 red flowers and 4 white flowers.

How many flowers are there altogether?

Let's draw a picture of the situation before you go to the next page.

① Which picture shows the situation, A or B?

A

B

② Let's write a math sentence and find the answer.

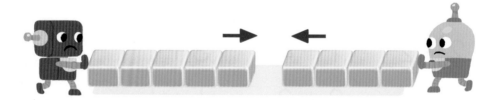

Math Sentence :

Answer : ☐ flowers

4 There are 2 black cats and 5 white cats. How many cats are there altogether?

① Let's draw a picture.

Nanami's picture

Daiki's picture

What is a good point in Daiki's picture?

Yui

② Let's write a math sentence in your notebook and find the answer.

Daiki's notebook

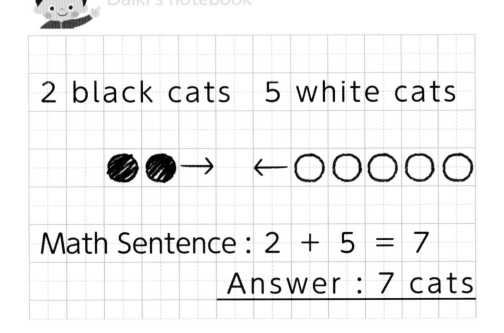

2 black cats   5 white cats

Math Sentence : 2 + 5 = 7

Answer : 7 cats

5 Let's make a math story for 5 + 3.

There are ⬚ monkeys.

There are ⬚ monkeys.

How many monkeys are there ⬚ ?

 3 Let's find the answers.

5 + 1   5 + 2   4 + 5   2 + 5

Let's look at the pictures and tell a story.

Putting 2 more goldfish ....

1 How many goldfish are there now after you put more?

Let's think about the problem by using blocks.

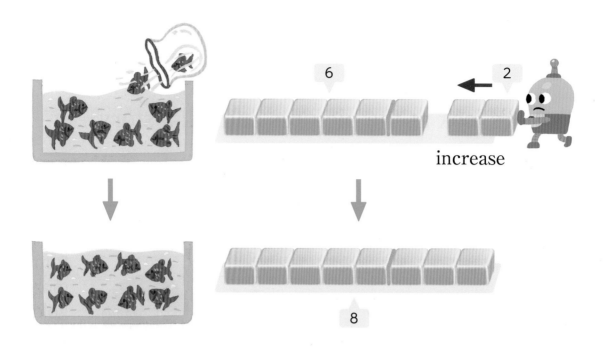

There are 6 goldfish. Putting 2 more goldfish,

it becomes 8 goldfish.

Math Sentence : 6 + 2 = 8

Answer : ☐ goldfish

2 How many candies are there altogether?

Math Sentence :

☐ + ☐ = ☐

Answer : ☐ candies

1 How many children are there altogether?

Math Sentence :

☐ + ☐ = ☐   Answer : ☐ children

2 Let's find the answers.

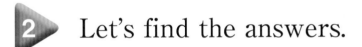

8 + 1   7 + 2   1 + 6   2 + 6

3 There are 4 cars at the parking lot. When 3 more cars come into the parking lot, how many cars are there altogether?

Let's draw a picture of the situation before you go to the next page.

① Which picture shows the situation, A or B?

A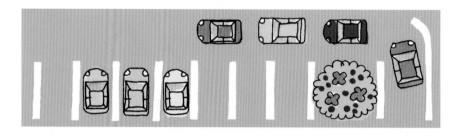

B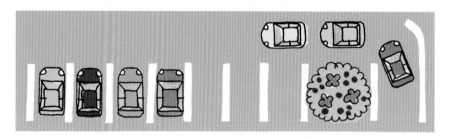

② Let's write a math sentence and find

the answer.

Math Sentence :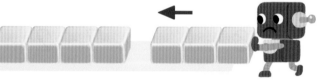

| |

Answer : | | cars

4 You have 4 pencils.

Mother gives you 2 more pencils.

How many pencils do you have

altogether?

① Let's draw a picture.

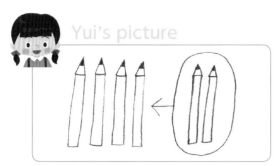

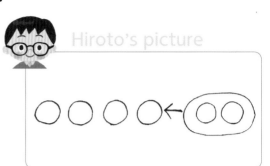

② Let's write a math sentence in

your notebook and find the answer.

It is easier to write by using ○.

Hiroto's notebook

| At first I had 4 pencils. | Mother gives me 2 pencils. |

○○○○ ←(○○)

Math Sentence : 4 + 2 = 6

Answer : 6 pencils

 3 Let's find the answers.

2 + 4   3 + 3   3 + 4   4 + 4

5 | Let's make a math story for 6 + 4.

There are ☐ cats.

☐ more cats come in.

How many cats are there  ?

▶ 4 Let's make a math story for 3 + 7.

▶ 5 Let's find the answers.

9 + 1   5 + 5   4 + 6   2 + 8

7 + 3   8 + 2   1 + 9   3 + 7

Addition Cards

card
front **4 + 3**
back **7**

Let's make addition cards and practice addition facts.

 1 Say the answer.

2 Find other cards with the same answer.

3 Line up the cards with the same answer in order.

You have two turns to throw balls into a basket. How many balls are there in the basket altogether?

Yui

first turn   second turn

2 + 1 = ☐

Hiroto

first turn   second turn

2 + ☐ = ☐

Nanami

first turn   second turn

☐ + ☐ = ☐

1 ▶ After throwing balls, 0 + 4 is written as the math expression. How many balls were inside the basket on each turn?

2 ▶ Let's find the answers.

4 + 0   9 + 0   7 + 0   8 + 0

0 + 6   0 + 5   0 + 1   0 + 0

Picture Book for Addition

Let's make a story for addition.

Together

Book of 4 + 5 ①

There are 4 blue cups.

There are 5 red cups.

④ 4 + 5 = 9

There are 9 cups altogether.

Increase

Book of 3 + 2 ①

There are 3 bcctlcs.

2 more beetles fly in.

④ 3 + 2 = 5

There are 5 beetles altogether.

What you can do now

☐ Can do addition.

1 Let's find the answers.

3 + 5	0 + 3	6 + 1	2 + 6
2 + 7	1 + 5	5 + 4	3 + 3
5 + 0	2 + 4	6 + 4	7 + 1
6 + 0	8 + 2	1 + 7	3 + 4

☐ Can find the addition expressions with the same answer.

2 Let's connect the cards with the same answer.

3 + 5 •	• 4 + 4
2 + 4 •	• 6 + 3
4 + 5 •	• 5 + 1

☐ Can make an addition expression and find the answer.

3 There are 6 candies.

You get 2 more candies.

How many candies do you have altogether?

Let's Play a Rock-Paper-Scissors Game

 If you win by paper,
you take 2 steps forward.

 If you win by rock,
you take 4 steps forward.

 If you win by scissors,
you take 6 steps forward.

Daiki

I won twice and took 6 steps forward.
Can you guess what I did?

Yui

I want to take 10 steps forward. What should I do?
Should I win by using rock, paper or scissors? And how many times?

If you win by scissors, you can take 6 steps forward.
How many steps are needed for 10 steps altogether?

4 How Many : Left and Difference

Let's look at the pictures and tell a story.

①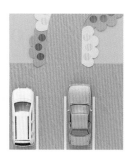

There are ☐ cars. ☐ cars go away. ☐ cars are left.

②

There are ☐ blocks. You take ☐ block. ☐ blocks are left.

How many goldfish are left?

Let's think about the problem by using blocks.

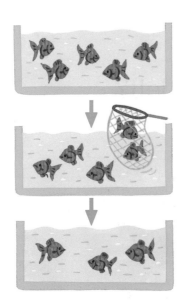
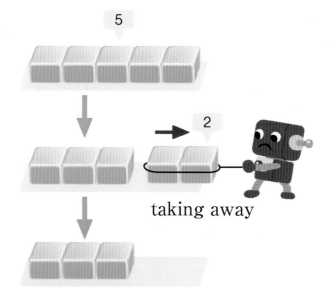

taking away

Taking away 2 from 5 is 3.

Math Sentence : 5 − 2 = 3

5 minus 2 equals 3.

Answer : 3 goldfish

5 − 2 = 3 is a math sentence.
5 − 2 is a math expression.

 2 How many are left?

I ate one piece.

Math Sentence :

☐ − ☐ = ☐

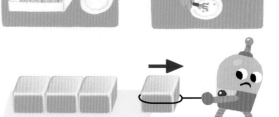

Answer : ☐ pieces

1 How many flowers are left?

 I gave 2 flowers.

Math Sentence :

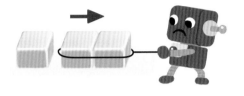

□ − □ = □

Answer : □ flowers

2 Let's do **Subtraction**.

5 − 3   2 − 1   4 − 2   5 − 4

4 − 3   3 − 1   5 − 1   5 − 2

3 There were 9 sheets of origami paper.

You used 4 sheets to make paper airplanes.

How many sheets were left?

Let's draw a picture of the situation.

① Which picture shows the situation, A or B?

A

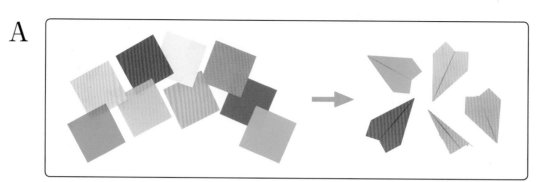

B

② Let's write a math sentence and find the answer.

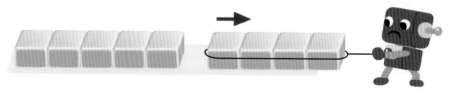

Math Sentence :

Answer : ☐ sheets

 Let's find the answers.

7 − 2   8 − 3   6 − 5   8 − 5

4 There were 9 children playing.

After a while, 3 children went home.

How many children were still playing?

① Let's draw a picture.

Hiroto's picture

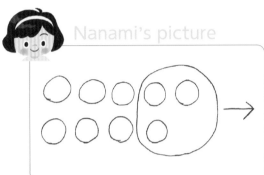

Nanami's picture

② Let's write a math sentence in your notebook and find the answer.

Nanami's notebook

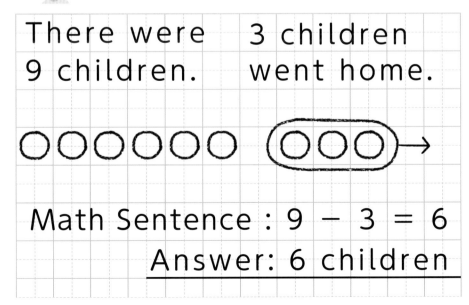

There were 3 children
9 children. went home.

Math Sentence : 9 − 3 = 6

Answer: 6 children

4 Let's find the answers.

$9 - 1$ $8 - 2$ $7 - 1$ $9 - 2$

5 Let's make a math story for $8 - 6$.

There were ☐ swallows on a wire.

☐ swallows flew away.

How many swallows were ☐ ?

 5 Let's find the answers.

$9 - 7$ $7 - 6$ $6 - 3$ $8 - 7$

$6 - 4$ $7 - 3$ $9 - 6$ $7 - 4$

6 There are 10 lions. 6 of them are males. How many are females?

Math Sentence :

Answer : lions

6 There were 10 pencils. He sharpened 3 of them. How many pencils were unsharpened?

7 Let's find the answers.

10 − 4   10 − 1   10 − 9   10 − 2

10 − 6   10 − 8   10 − 7   10 − 5

Subtraction Cards

Let's make subtraction cards and practice subtraction facts.

card

front $7-2$

back 5

⭐1 Say the answer.

⭐2 Find other cards with the same answer.

⭐3 Line up the cards with the same answer in order.

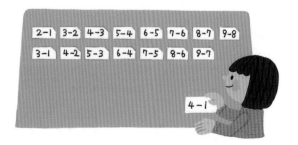

How many goldfish are left?

①  Taking away 1   3 − 1 = ☐

②  Taking away 2   3 − ☐ = ☐

③  Taking away 3   3 − ☐ = ☐

④  Taking away 0   3 − 0 = ☐

 Let's find the answers.

7 − 7   4 − 4   5 − 5   9 − 9

8 − 0   1 − 0   6 − 0   0 − 0

How many difference?

1 How many more cows are there than horses?

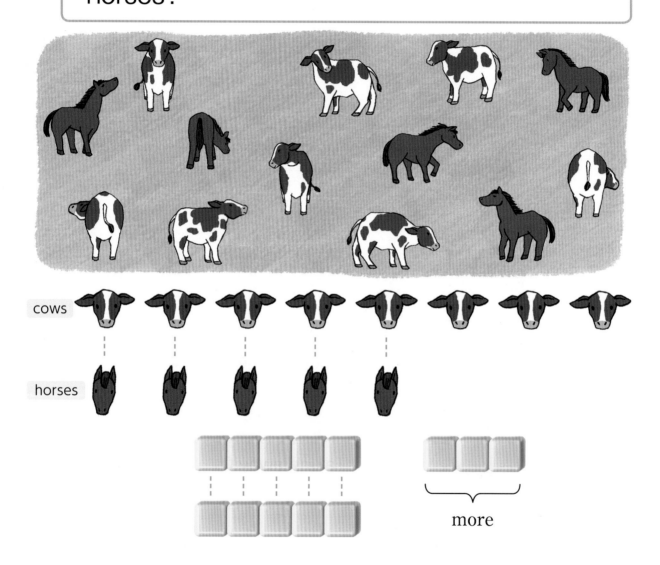

cows

horses

more

8 is 3 more than 5.

Math Sentence : 8 − 5 = ☐

Answer : ☐ more cows

1 How many more buses are there than trucks?

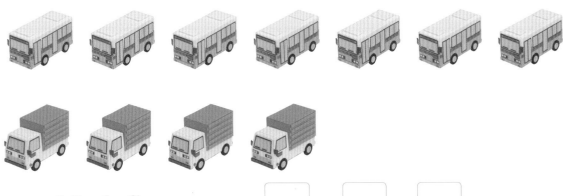

Math Sentence : ☐ − ☐ = ☐

Answer : ☐ more buses

2 How many fewer melons are there than watermelons?

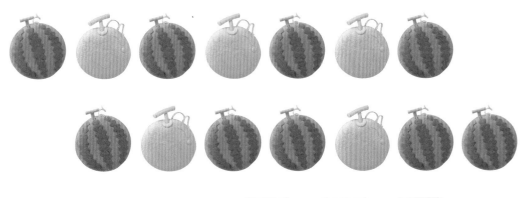

Math Sentence : ☐ − ☐ = ☐

Answer : ☐ fewer melons

2 There are red cars and yellow cars.

Which is more and by how many?

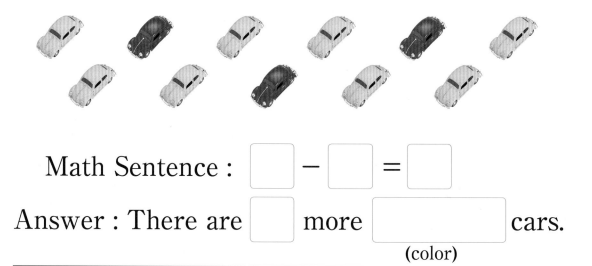

Math Sentence : ☐ − ☐ = ☐

Answer : There are ☐ more ☐ cars.

(color)

---

3 There are cats and dogs.

Which is fewer and by how many?

Math Sentence : ☐ − ☐ = ☐

Answer : There are ☐ fewer ☐ .

3 What is the difference between the number of children and the number of candies?

Math Sentence :

Answer :

4 What is the difference between the number of plates and the number of pieces of cake?

Math Sentence :

Answer :

Picture Book for Subtraction

Let's make a story for subtraction.

Left

Book of 6 − 2 ①

There were 6 bananas.

He ate 2 bananas.

④ 6 − 2 = 4

There are 4 bananas left.

Difference

Book of 4 − 3 ①

There are 4 oranges.

There are 3 apples.

④ 4 − 3 = 1

There is 1 more orange than apples.

What you can do now

☐ Can do subtraction.

1 Let's find the answers.

4 – 1	9 – 8	2 – 2
6 – 2	7 – 5	8 – 8
7 – 0	10 – 3	8 – 1

☐ Can make a subtraction expression and find the answer.

2 Let's write a math expression and find the answer.

① There were 8 apples.

You ate 4 apples.

How many apples were left?

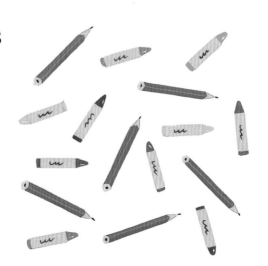

② There are 6 pencils and 10 crayons.

Which is more and by how many?

5 Ordinal Numbers

The first **4** children

The **4**th child

Let's color it!

The first **2** cars

The **2**nd car

The **3**rd car from the end of line

1 What is the position of the sheep?

It's the ☐ from the top.

Daiki

It's the ☐ from the bottom.

Nanami

 What is the position of the monkey?

It's the ☐ from the right.

Yui

It's the ☐ from the left.

Hiroto

2 Let's raise the hand when your position is called.

The 3rd student.

The first 3 students.

2 Let's find other cards with the same picture.

There is a strawberry on the 5th card from the left.

6 How Many in Each Group

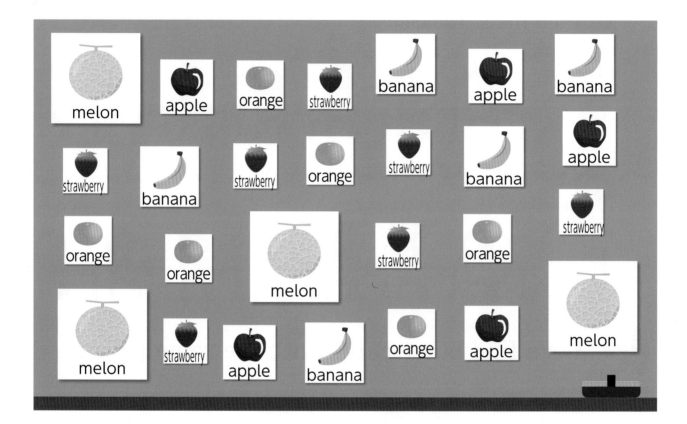

1 Let's examine the number of fruits.

How shall I count them?

Nanami

I'll try to line up them.

Yui

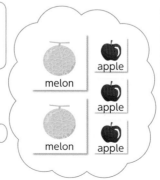

But they don't have the same size.

Daiki

① Let's think about the ways of counting with your friends.

② Let's color the number of fruits in each group.

③ Which group of fruits has the largest number?

④ What is the fruit of which number is 6?

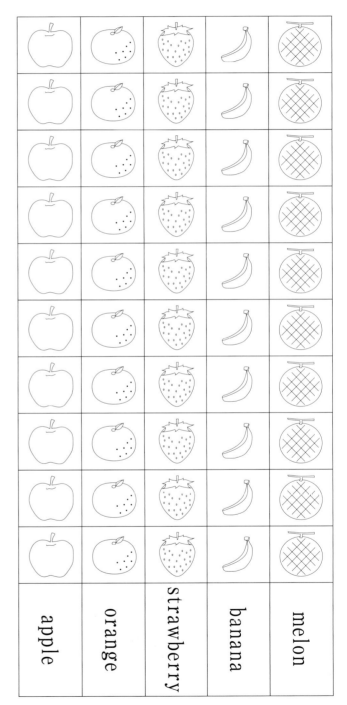

Look around the classroom. Let's make a lot of addition and subtraction expressions and create a story.

Want to explore   In the classroom

You will do group work in math class.

Everyone looks like they are having fun.

1 Let's make a lot of stories and expressions of addition and subtraction, and then tell them to your friends.

> Can you make the same story and expression as your friend's?
> Can you make a different story and expression from your friend's?

2 Let's try to make stories and expressions of addition and subtraction with the tools on the desks shown on the left page.

Yui

> There were 5 pencils on the desk of Group 1.
> 2 pencils were handed to Mei of Group 2.
> How many pencils were left?

> 3 pencils were left.

Daiki

3 Let's make a presentation of your story.

> First of all, the order of presentations should be decided.
> What is your presentation turn?
> Now let's give a presentation in order.

01101

7 Numbers Larger than 10

Numbers up to 20

1 Let's count the number of squirrels.

Yui
I'll try to use blocks.

After counting 10, how many squirrels are left?

Daiki

74

Let's count the number of acorns.

 Let's count the number of walnuts.

10 and 3 make...

10 and 3

13 squirrels

Daiki

10 and 5 make...

10 and 5

☐ acorns

Yui

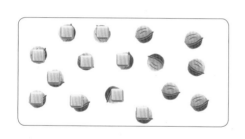

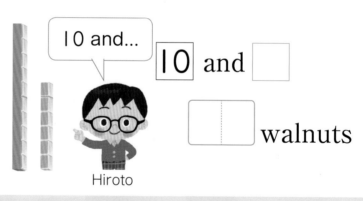

10 and...

10 and ☐

☐ walnuts

Hiroto

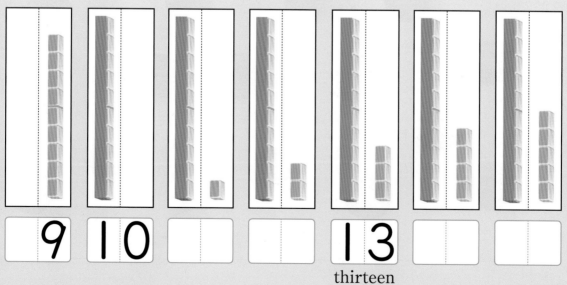

9   10   ☐   ☐   13   ☐   ☐

thirteen

2 How many apples and eggs are there?

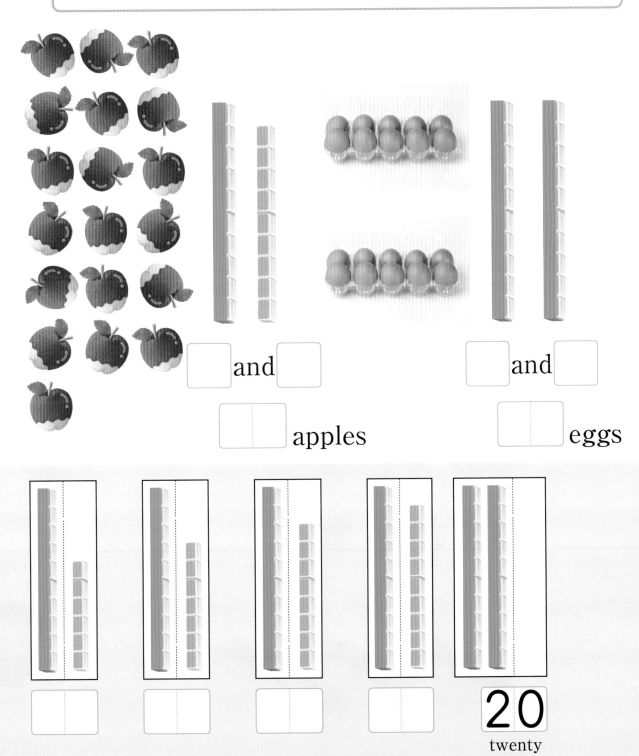

⬜ and ⬜   ⬜ and ⬜

⬜⬜ apples   ⬜⬜ eggs

20
twenty

3 Let's count the things below.

①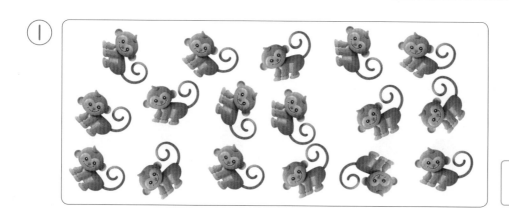

monkeys

②

owls

③

two, four, six, eight, ...

Nanami

strawberries

④

five, ten, ...

Daiki

pieces

4 Let's fill in each ☐ with a number.

① 10 and 2 make ☐.

② 10 and 5 make ☐.

③ 10 and 8 make ☐.

Can we use blocks?

Nanami

3 Let's fill in each ☐ with a number.

① 13 is ☐ and 3.

② 16 is 10 and ☐.

③ 19 is 10 and ☐.

4 Let's fill in each ☐ with a number.

① 10 and 4 make ☐.

② 10 and 7 make ☐.

③ 11 is ☐ and 1.

④ 20 is 10 and ☐.

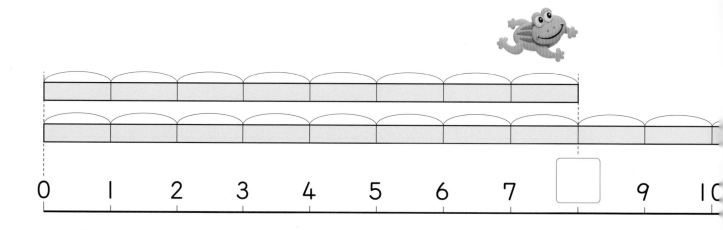

0 1 2 3 4 5 6 7 ☐ 9 1C

5 Let's find the answer by using the line of numbers above.

① How far did the jump?

② How far did the  jump?

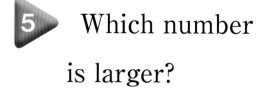 Which number is larger?

On the line of numbers, the number becomes larger as you move towards the right.

①  9 11

② 15 13

③ 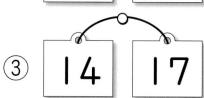 14 17

④ 20 18

| 11 | 12 | 13 | 14 | 15 | 16 | | 18 | 19 | 20 |

6 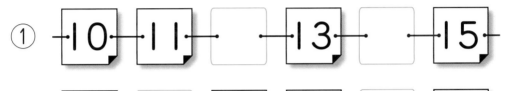 Let's fill in each ☐ with a number.

① ⌐10⌐⌐11⌐⌐ ☐ ⌐⌐13⌐⌐ ☐ ⌐⌐15⌐

② ⌐15⌐⌐ ☐ ⌐⌐17⌐⌐18⌐⌐ ☐ ⌐⌐20⌐

③ ⌐ ☐ ⌐⌐19⌐⌐18⌐⌐17⌐⌐ ☐ ⌐⌐15⌐

7 Let's fill in each ☐ with a number.

① 3 larger than 12 is ☐ .

② 4 smaller than 18 is ☐ .

Want to connect

There are numbers larger than 20, right?

Daiki

13 is 10 and 3.

Let's fill in each ☐ with a number.

① The number when you add 3 to 10.

$$10 + 3 = \boxed{}$$

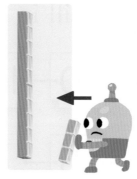

② The number when you subtract 3 from 13.

$$13 - 3 = \boxed{}$$

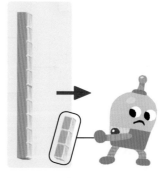

Addition and subtraction are called calculation or operation.

 Let's do the **calculation**.

$10 + 2$	$10 + 4$	$10 + 7$	$10 + 9$
$12 - 2$	$15 - 5$	$16 - 6$	$19 - 9$

2 There are 12 candies.

When you get 3 more candies,

how many do you have altogether?

Math Sentence:

☐ + ☐ = ☐

Answer : ☐ candies

2 ▶ There are 15 tomatoes.

When you eat 2 tomatoes,

how many are left?

Math Sentence:

☐ − ☐ = ☐

Answer : ☐ tomatoes

3 ▶ Let's calculate.

| 11 + 4 | 13 + 4 | 16 + 3 | 12 + 6 |

| 14 − 1 | 17 − 2 | 18 − 5 | 16 − 4 |

8 Time (1)

1 Let's look at the above pictures and tell a story.

Telling Time

long hand

short hand

The short hand is at 8 and the long hand is at 12, so it's 8 o'clock.

The short hand is between 8 and 9, and the long hand is at 6, so it's half past 8.

Let's think by
using the clock
on p.106.

2 Let's show the time by moving the hands of the clock.

① 11 o'clock   ② half past 3

1 ▶ Let's draw the long hand on the clock to show the time.

① 6 o'clock

② half past 4

③ half past 11

85

9 Shapes (1)

1 Let's collect various shapes.

Let's make similar groups.

2 Let's play with shapes.

Let's roll the shapes.

A ball rolls well.

Some shapes roll, and others don't.

Let's stack the shapes.

Some shapes cannot be stacked.

3 Let's guess the shapes in the box.

Let's explain why you think so.

There are no flat parts, so ....

A B C D

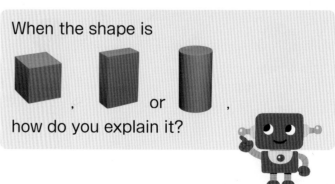

When the shape is

, or ,

how do you explain it?

Let's build various things you like.

I use this box for an arm.

A round shape is good for a tire.

Let's make a presentation about your work.

What did you build?

I built an airplane with long wings.

Let's draw a picture by tracing shapes.

Let's try to cut them out after tracing.

I'm using this triangle to draw a roof.

Let's make a presentation about your picture.

1 What is the shape by tracing on the paper?

Let's connect the points with lines.

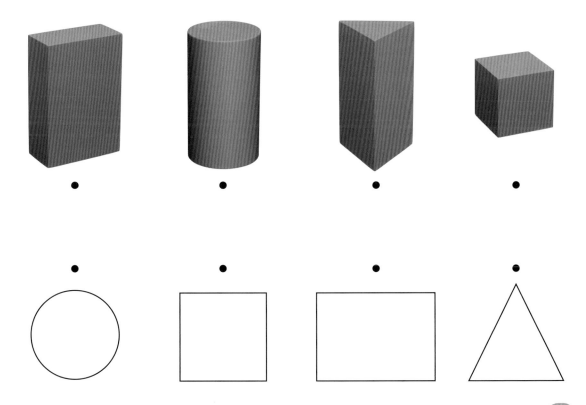

10 Addition or Subtraction

1 How many people are there altogether?

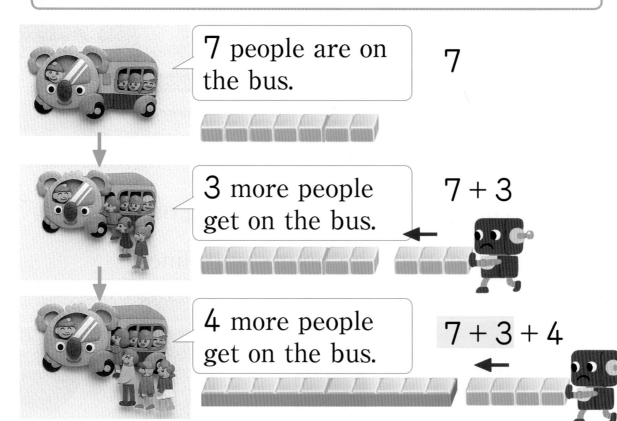

7 people are on the bus.   7

3 more people get on the bus.   7 + 3

4 more people get on the bus.   7 + 3 + 4

Math Sentence : $7 + 3 + 4 = \boxed{}$

Answer : $\boxed{}$ people

 1 Let's do the calculation.

$$8 + 2 + 3 \qquad 5 + 5 + 7 \qquad 6 + 4 + 9$$

2 How many strawberries are left?

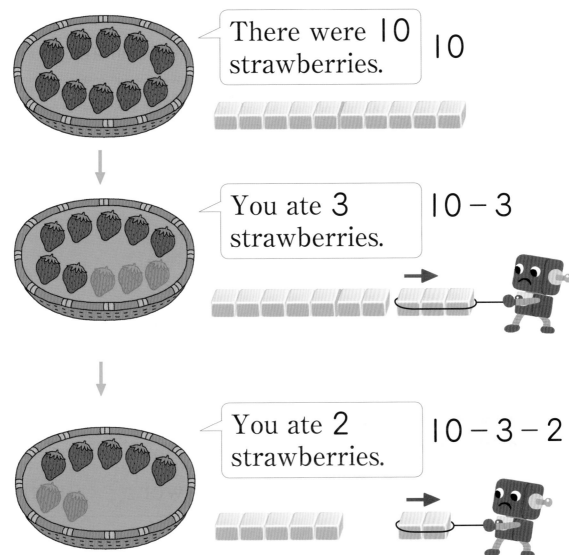

There were 10 strawberries.   10

You ate 3 strawberries.   10 − 3

You ate 2 strawberries.   10 − 3 − 2

Math Sentence : 10 − 3 − 2 = ☐

Answer : ☐ strawberries

 3 Let's calculate.

$$10 - 5 - 1 \qquad 10 - 3 - 4 \qquad 12 - 2 - 6$$

2 There are 10 children.

6 children go home, and 3 more children come.

How many children are there now?

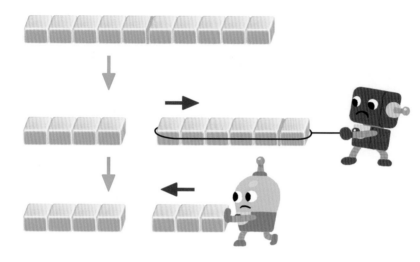

10

10 − 6

10 − 6 + 3

Math Sentence : 10 − 6 + 3 = ☐

Answer : ☐ children

 4 Let's make a math story for 5 + 2 − 3.

 5 Let's calculate.

8 − 5 + 4   10 − 7 + 2   13 − 3 + 5

5 + 2 − 1   4 + 6 − 7   5 + 5 − 9

Supplementary Problems

1 Numbers up to 10

pp.6~23

1 Let's write the number of dots.

①

②

③

④

2 Which is more?

①

②

3 Which number is larger?

① 4 3

② 2 5

③ 6 9

④ 1 0

4 Let's write the number in the ☐.

① — 3 — ☐ — 5 —

② — ☐ — 7 — 8 —

③ — 0 — ☐ — 2 —

④ — 8 — 9 — ☐ —

2 Decomposing and Composing Numbers

pp.24~31

1 Let's write the number in the ☐.

① ●●○○○
5 is 2 and ☐.

② ●●●●○
5 is 4 and ☐.

③ ●●○○○○
6 is 2 and ☐.

④ ●●●○○○
6 is 3 and ☐.

⑤ ●●●●●○○○
7 is 4 and ☐.

⑥ ●●○○○○○
7 is 2 and ☐.

⑦ ●●●●●●○
7 is 6 and ☐.

2 8 and 9 each are decomposed into 2 numbers.

Let's write the number in the ☐.

①
8	
3	

②
8	
	2

③
8	
1	

④
8	
	4

⑤
9	
7	

⑥
9	
	6

⑦
9	
	5

⑧
9	
8	

3 Let's connect the two numbers which make 10.

3	6	8	5	1

5	2	7	9	4

97

③ How Many : Altogether and Increase

pp.32~49

1 There are 2 red flowers and 3 white flowers.

How many flowers are there altogether?

Math Sentence :

Answer : flowers

2 You have 3 balloons.

You get 1 more balloon.

How many balloons do you have altogether?

Math Sentence :

Answer : balloons

3 Let's find the answers.

① 3 + 2   ② 1 + 1

③ 2 + 2   ④ 2 + 1

⑤ 1 + 4   ⑥ 1 + 3

4 There are 4 boys and 3 girls.

How many children are there altogether?

5 There are 6 cars at the parking lot.

When 3 more cars come into the parking lot, how many cars are there altogether?

6 Let's find the answers.

① $2+5$ ② $7+1$

③ $3+6$ ④ $5+3$

⑤ $1+8$ ⑥ $4+4$

⑦ $2+7$ ⑧ $3+5$

7 There are 6 sheets of red origami paper and 4 sheets of blue origami paper.

How many sheets are there altogether?

8 Let's find the answers.

① $2+8$ ② $1+9$

③ $3+7$ ④ $5+5$

⑤ $4+6$ ⑥ $7+3$

⑦ $8+0$ ⑧ $3+0$

⑨ $0+6$ ⑩ $0+0$

4 How Many : Left and Difference

pp.50~65

1 There are 7 goldfish.

You take away 2 goldfish.

How many goldfish are left?

Math Sentence :

[   ]

Answer : [   ] goldfish

2 There were 6 apples.

You ate 2 apples.

How many apples are left?

Math Sentence :

[   ]

Answer : [   ] apples

3 Let's find the answers.

① $4 - 3$ ② $9 - 3$

③ $8 - 5$ ④ $5 - 1$

⑤ $7 - 3$ ⑥ $3 - 2$

⑦ $6 - 5$ ⑧ $8 - 6$

⑨ $6 - 3$ ⑩ $7 - 2$

4 There were 10 flags.
You took 7 flags.
How many flags are left?

5 Let's find the answers.

① $10 - 3$ ② $10 - 1$

③ $10 - 5$ ④ $10 - 2$

⑤ $10 - 4$ ⑥ $10 - 9$

⑦ $6 - 6$ ⑧ $1 - 1$

⑨ $3 - 3$ ⑩ $9 - 0$

⑪ $2 - 0$ ⑫ $4 - 0$

6 There are 6 oranges and 4 apples.
What is the difference between the number of oranges and the number of apples?

7 There are 7 boys and 9 girls.
Which is more and by how many?

8 Connect the math expressions with the same answer.

$5 - 3$ •	• $9 - 6$
$6 - 2$ •	• $5 - 0$
$7 - 4$ •	• $7 - 1$
$8 - 2$ •	• $10 - 8$
$9 - 4$ •	• $8 - 4$

❺ Ordinal Numbers

pp.66~69

Ⅰ Let's color it.

① The first **3** circles from the left

left  right

② The **3rd** circle from the left

left  right

③ The first **2** circles from the right

left  right

④ The **2nd** circle from the right

left  right

2 Children are lined up.

front Haruto Hina Mei Ren Sota Yua back

① Who is the **3rd** child from the front?

② What is the position of Ren from the front?

③ Who is the **2nd** child from the back?

④ What is the position of Mei from the back?

⑤ Who is the **2nd** child from the front? What is his/her position from the back?

⑦ Numbers Larger than 10

pp.74~83

1 Let's count the number of blocks.

①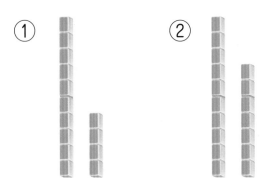
②

2 Let's write the number in the ☐ .

① 10 and 3 make ☐ .

② 10 and 6 make ☐ .

③ 11 is ☐ and 10.

④ 20 is ☐ and 10.

⑤ 15 is 10 and ☐ .

⑥ 18 is 10 and ☐ .

3 Which number is larger?

①
| 12 | 10 |

②
| 15 | 17 |

③
| 19 | 16 |

4 Let's write the number in the ☐ .

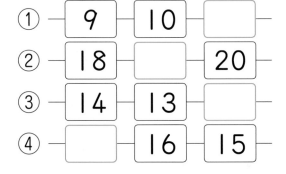

① 9 — 10 — ☐

② 18 — ☐ — 20

③ 14 — 13 — ☐

④ ☐ — 16 — 15

5 Let's find the answers.

① $10 + 5$ ② $10 + 8$

③ $14 - 4$ ④ $18 - 8$

6 Let's find the answers.

① $12 + 5$ ② $11 + 6$

③ $14 + 4$ ④ $17 + 2$

⑤ $13 - 2$ ⑥ $15 - 4$

⑦ $17 - 3$ ⑧ $19 - 5$

7 There are 11 sheets of origami paper.

8 more sheets are given.

How many sheets are there altogether?

8 There were 16 candies.

You gave someone 5 candies.

How many candies are left?

8 Time (1)

pp.84~85

1 What time is it?

①

②

③

④

2 Draw the long hand on the clock to show the time.

① 5 o'clock

② 11 o'clock

③ half past 2

④ half past 9

9 Shapes (1)
pp.86~91

1 Let's connect the shapes that are the same.

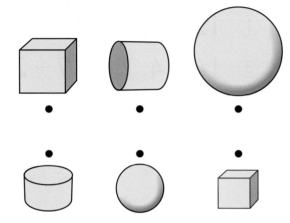

2 Let's connect the shapes with their tracing.

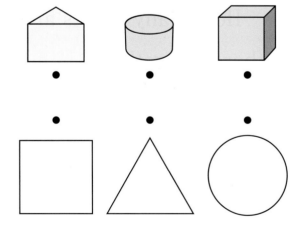

⑩ Addition or Subtraction

pp.92~94

1 Let's find the answers.

① $4 + 6 + 3$

② $9 + 1 + 7$

③ $3 + 7 + 5$

2 There were 8 cars at the parking lot.

2 more cars came there, and then 3 more cars came.

How many cars were there altogether?

3 Let's find the answers.

① $9 - 3 - 2$

② $8 - 4 - 1$

③ $10 - 6 - 3$

4 There were 7 oranges.

Kento ate 2 oranges and Yumi ate 3.

How many oranges are left?

5 Let's find the answers.

① $10 - 6 + 2$

② $10 - 8 + 4$

③ $6 + 3 - 5$

6 There were 10 people on the bus.

At the bus stop, 5 people got off the bus and 3 people got on.

How many people are there on the bus now?

Clock ▼ will be used in pages 84 and 85.

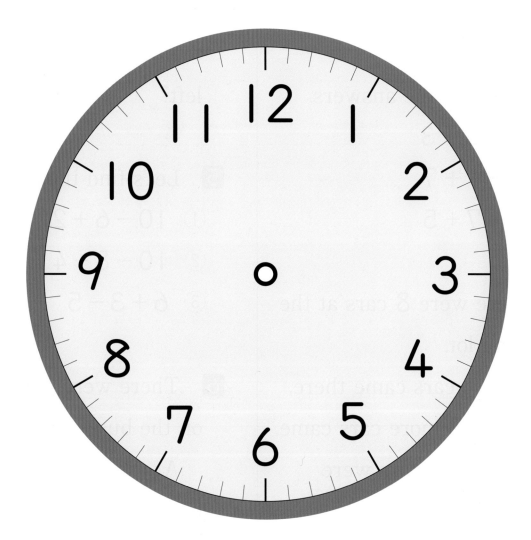

Editorial for English Edition:

Study with Your Friends, Mathematics for Elementary School
1st Grade, Vol.1, Gakko Tosho Co.,Ltd., Tokyo, Japan [2020]